I0390642

Adaptation of the Rat Femur to a Force of 2g

Wilson P. McCord Jr.

1993

PUBLISHING HISTORY

Adaptation of the Rat Femur to a Force of 2g by Wilson P. McCord Jr. © 1993. Digital Publication, 2013 Wilson McCord Science Publication, New York, New York. All rights reserved. All Digital Rights belong to Wilson McCord, 2013. Text, Photographs and Illustrations Wilson P. McCord Jr. Copyright © 1993

📖

The Masters Thesis, "Adaptation of the Rat Femur to a Force of 2g" was Submitted by Wilson McCord in partial fulfillment of the requirements for the degree of Master of Anthropology, Hunter College, The City University of New York and Published on University Microfilms International, May 26, 1993. Copyright © 1993 by UMI and Hunter College of the City University of New York. All rights reserved Text / Illustration Copyright © 1993 Wilson P. McCord Jr.

📖

Cover and Frontis Piece Photographs of Earth Courtesy of NASA Cover design by Wilson McCord © 2013 Wilson McCord and the WM eye colophon are trademarks of Wilson McCord

📖

All rights reserved under international copyright conventions. No part of this book may be reproduced or utilized in any form or by any means, electronic or mechanical including photocopying, recording on any information or retrieval system without permission in writing from the publisher.

WILSON MCCORD SCIENCE PUBLICATIONS

Adaptation of the Rat Femur to a Force of 2g

Wilson P. McCord Jr., M.A.

Submitted in partial fulfillment of the

requirements for the degree of

Master of Anthropology

Hunter College, The City University of New York

In cooperation with AMES Research Center

NASA

26 May 1993

4

Acknowledgements

All of the research in this paper was done in the Biological Anthropology Section, Department of Anthropology, Hunter College of The City University of New York, NY 10021.

Preface

The project set out to determine if changes during loading, in this case living at 2g, in bone is on the tissue or biochemical level and whether the morphological character of collagen fiber orientation changes due to this environmental stress. The project was selected with the goal of contributing to the reverse engineering of bone. The research done in this Thesis is one of the earliest, if not the first, examples of bioengineering bone. The images here have been digitally enhanced, made clearer and bubbles were removed in the medium for this digital publication. Measurement bars were made clearer and solid for this publication for easier viewing. All original research: writing (except in the case of spelling corrections), photographs of the specimens themselves, schematics and original tables have been preserved from the original 1993 Thesis.

Wilson McCord, September 2013

Adaptation of the Rat Femur to a Force of 2g

Wilson McCord, 1993

Abstract

Adaptation to microgravity remains poorly understood. The present study analyzes the change in the mid-diaphysis and distal sections of the Rat femur after exposure to a force of 2g. The morphological structure of the femur is within the findings of recent spaceflight experiments. However there is a visible increase in the number of osteons in some of the specimens. These visible features signal the adaptation of the rat femur to the stress situation.

Introduction

Growth and functional adaptation help insure the development of normal bone architecture. The purpose for bone mass and the arrangement of its matrices is "the mechanical one" (Lanyon, p. s61, 84). Areas of the skeleton that are required to withstand stress must be technically able to cope with the mechanical stimuli that varies during life and changes with environment. The technical ability of this working tissue is its capability to influence remodeling (Lanyon, p. 1092, 87).

During the growth and development of all living things limitations on phenotypic variation are "caused by the structure, character, composition, or dynamics of the developmental system." These limitations have been defined by Smith et al. as developmental constraints (Smith et al. p. 265, 85).

Developmental constraints govern collagen orientation during growth. Later this architecture serves as the dynamic structure in which intracortical and surface remodeling occurs. Every bone in a given animal's skeletal system has its own construction rules as well as its own remodeling regime.

The same genetic program that sets the developmental constraints and makes the bone also programs a "Memory" for the set strain of any particular bone. It is present during growth and development and continues in maturity once growth has finished. This memory allows an adaptive response (remodeling of bone) to the mechanical stresses of the environment.

Bones can grow and develop without activating this memory by functional load bearing. But "... despite the structural suitability of the genetically determined arrangement, a bone that has developed in the absence of load bearing would fail if immediately subjected to normal functional loads" (Lanyon, S370, 92).

The features upon which load bearing competence depends are "... girth, cortical thickness, medullary cavity diameter, cross-sectional shape, and the orientation and spatial arrangement of trabeculae..." (Lanyon, S370, 92). All of these features "... develop "normally" and are maintained "normal only in the presence of functional load bearing" (Jaworski Z.6., 86). Bone architecture could not adapt or maintain its structural competence (bone remodeling) during changes in environment (functional requirements) without a strain-related stimulus.

Skerry et al. believe that the reorientation of proteoglycan provides a physical basis for a "Strain memory" in bone tissue (Skerry et al. 88). In short periods of loading they observed large changes in bone mass. If strain memory were involved a short period of loading could provide a stimulus continuing for 24h.

There is a coordination of growth remodeling, secondary Haversian remodeling and the infilling of cancellous spaces during the production of collagen and its orientation (Bromage, pp. 42-43, 89b). Biewener et al.'s studies have strongly supported "... the hypothesis that bones remodel or model during growth to maintain a uniform distribution of strains matched to a level of physical exercise to which they are subjected" (Biewener et al., pp. 392-3, 86). Here again we see evidence that there is a strain memory which responds to stress and influences adaptive remodeling.

The secondary Haversian remodeling serves to acquire certain fixed bone densities and constantly readjusts the bone mechanics in order to meet the physical laws or strain memory within the bones (Bromage, p.44, 89a). Haversian remodeling reacts to the strain history and redistributes the stress stimuli so changes in the surface adaptive remodeling will not occur and the interior geometry and size of the bone will stay intact (Bromage, p. 41, 89b).

Lanyon has studied the components of a dynamic strain regime which influences remodeling behavior (Lanyon, p.

1092, 87). In a single period of loading, bone remodeling response appears to saturate in only a few loading cycles, additional strain produced no extra effects (Lanyon, p. 1092, 87).

Exercise experiments have shown that functionally deprived bone, which has not carried loads for a long time; acquire low bone masses associated with disuse atrophy (Uhthoff p. s59, 84). The initial feature, during rest, of each remodeling event is resorptive. And the number of remodeling events is greater than those in bones that are enduring stress. The remodeling of the bones that are load bearing are in balance because their bone mass is constant (Lanyon, p. 59, 84).

Though their remodeling activities will be in balance, bones experiencing optimum strains, during loading cycles, will have a higher bone mass (Lanyon, p. s59, 84). Bones strained above this optimum level increase in bone mass. The "...strain-generated osteogenic stimulus..." "...ensures the appropriate bone mass for each manner and level of functional loading achieved" (Lanyon, s59, 84).

Similarly, when bone mass is unnecessarily high and the level of functional strain has been reduced, at 1g, the effect is absorption and, though a certain amount of new bone will be deposited it will not be supplemented by the strain generated osteogenic stimulus leaving the remodeling balance negative. "This is the typical osteoporosis reaction by which bone loss is normally achieved" (Lanyon, p. s60, 84). Hence, osteoporosis can be induced by immobilization.

'Secondary Haversian' (intracortical) remodeling is the processes responsible for the structural stability of bone tissues under functional loading. It is this system which maintains and repairs the tissue.

The failure of newly formed bone collagen to acquire mineral is said to be the principle effect on bone during spaceflight (Mechanic et al., p.34, 90). Bone measurements of young growing rats taken on a seven day space shuttle mission and two Soviet biosatellite flights of twelve and 18.5 days revealed an underdevelopment of hard tissues. The time period required for these matrices to mature was also lengthened considerably. Previously existing trabecular bone structure was lost through resorption without an increase in osteoclasts or activity. All of these symptoms lead to a decrease in the biomechanical strength of the weight bearing parts of the rat skeleton (D. J. Simmons et al., p. 29, 90).

Rats, like all vertebrate groups, have a basic complement of structures or elements that differ from other vertebrates only in the combination and organization of these fundamental complement of structures (Enlow and Brown, p. 212, 58). Thus it is possible to observe bone growth of these animals in space and use the data to hypothesize what growth and adaptation would be like for humans living under microgravitational conditions.

While earlier tests have shown a change in bone mineral density, Bromage has noted, in examining the threads of collagen in rats formed during spaceflight, that there is no difference between the seams of collagen during the growth of rats flown in space and the control group's collagen alignment here on earth.

While the emphasis on collagen fiber direction

continues to be an important factor in a bone's ability to cope with stress, it has been shown that the absence of strain does not effect collagen fiber orientation. This finding coincides with Jaworski's work described above. But the mineral content does decrease and it can be said that the environment of space (microgravity) stimulates the adaptive system allowing bones to reset the mineral requirement of newly formed bone collagen in coordination with their strain memory. This adaptive response of bone is a key issue in the environment of space.

For many years astronauts have been exercising in space, but the effects of weightlessness require that, the longer the flight duration the longer the exercise regime. Still, the microgravity conditions diminish the osteogenic stimulus.

"Immobility due to fracture, joint disease, hemiplegia (Hodkinson and Brain, 1967), and the application of splints leads to localized osteoporosis." It has been demonstrated and noted by Donaldson et al. (1970) that continued bed rest will readily activate bone resorption causing disuse atrophy. But "... restoration of bone can occur to a considerable degree on remobilization. "A similar rapid demineralization of bone demonstrated in fit men subjected to microgravity on Skylab missions (Tilton et al., 80) prompted much of the experimentation done on bone during present day missions into space.

More recently it has been shown by Garetto, L.P. et al. that after spaceflight there is a strong and rapid recovery mechanism for osteoblast differentiation that is not suppressed by physiological stress. It is apparent from the strength of the osteogenic response in the face of chronic stress conditions that recovery of osteogenic potential is a physiological priority after return to a 1g environment.

Though the study of rat bone gives us enough data to determine the initial effects of space flight on bone tissue, in relation to the human skeleton, The 100 day old rats used in the Cosmos 2044 biosatellite experiments have no secondary Haversian remodeling system. Bromage notes that this is "...an important physiological and mechanical feature of human skeletal dynamics ..." and suggests the use of a larger mammal on future missions (Bromage, p. 6, 91).

Past observations of spaceflight animals have shown that exposure to launch and reentry forces, noise and vibration do not significantly aggravate or increase microgravity effects. But physiological responses to stimuli such as these in conjunction with microgravity environment are not known, nor will ideal spaceflight condition allow such exacting data (Vailas, p. 10, 90). Thus "... the changes observed in the rat bones cannot be solely attributed to the stress of landing or spaceflight" (Mechanic, p.34, 90). What can be said is, that the decreases in newly formed bone to acquire mineral will take place when humans experience spaceflight.

Monkeys, like humans, showed decrease bone mineral and evidence of increased bone resorption. In rats the primary reason for decreased bone mass is a reduced rate of formation without significant resorption. This decreased formation in turn leads to slower growth demineralization and decreased bending strength. The different response of bone in rats than in humans or monkeys could be due to a different type of bone structure or because in rats there is constant bone growth.

Simmons et al. (90) have recently affirmed the

13

concept that the microgravity not only affects the maturation of newly formed matrix and mineral moieties in weight-bearing bone, but, causes similar effects throughout the skeleton. In the past human populations have adapted to many physical environmental stresses. It is clear that the adaptive capacity of human populations to stress contains a number of response mechanisms. And though "... our ability to cope with the removal of stress, such as gravity is unclear..." (Harrison, p.478, 89) the extensive loading of bones and their immobilization show clear cut adaptability.

In the past biological adaptation has allowed survival in the ever changing environment. "We continue to spread far beyond the original habitat in which we evolved" (Harrison, p. 439, 89). And, a very big part of our adaptability has come from our ability to manipulate and almost create environments suitable to our needs. Thus NASA has created an environment of 2g to test the adaptive stimuli of bone so to observe the changes in the cortex of the femur of rats subjected to this environment.

Present Experiment

Cooper et al., 1966; Ascenzi et al., 1973; Ascenzi and Bannock, 1976 have used electron micrographs resulting in agreement with Gebhardt's (1906) theory that the compressive strength is greatest for osteons of transversal type and that tension strength is greatest for osteons of Longitudinal types.

Lamellae that are best able to support tensile stress (those with longitudinally oriented fiber bundles) are dark under the polarizing microscope in bone cross sections. Lamellae best able to support compression (those with transversely oriented fiber bundles) are bright under polarizing microscope in bone cross sections (Ascenzi, 1983).

Analysis of the distribution and occurrence of these two types of lamellae will tell us whether an osteon is better able to support tensile or compressive stress, and the difference between these two capabilities. Measurement of cortices, bone widths and areas and the occurrence of osteons (Haversian systems) and their densities will tell us more about the bone's strain related history.

Materials and Methods

Environment

Male specific pathogen-free, Wistar rats (n=10) were randomly assigned to two groups: synchronous control (S) and experiment (E). Each group was exposed to the same environmental conditions (weather) except that group (E) lived at 2g during the length of the experiment (2 weeks). Environmental conditions were set: air pressure 760mm Hg, humidity average 58% and ambient temperature 22-23 degrees Celsius.

The lighting was designed to 16h of light and 8h of dark with the light period beginning at 8:OOA.M. (0800 to 2400h lights on and from 2400 to 0800h lights off). Light intensity was (4-8) lumens at the cage floor, with an incandescent lamp place over each feeder.

Animals

The rat cages were 65X20X16 cm and had nozzles for delivery of their paste diet and lixits for water consumption. All rats were given a paste diet that is 70% water. ~14-g aliquots of paste diet are provided four times a day at 6h intervals starting at 2A.M. Food consumption averages ~45-55g for both groups. Water was provided to both groups. Water consumption, other than that contained in the diet, is typically ~2ml/day.

All rats in the NASA and Russian satellite experiments, including ours, are adapted to the diet, on average, from 2-3 weeks before the beginning of the experiments and were also adapted to the experimental cages at the same time. At the termination of the experiment both groups were killed on the same day.

Morphology

All specimens were obtained from AMES, NASA in California. All extraneous soft tissue was removed. The specimens were then sectioned for ultrasonic, morphological, and biochemical testing. A section of a femur from each rat was sent in separate bottles to our research unit. Each bottle contained a section of femur from the mid-diaphysis to the distal shaft of the femur.

Section preparation

Using an electric rotating saw (Buehler, ISOMET TM, Low Speed Saw) 350 microns in thickness cross-serial sections 100 microns thick were cut. The distance between two successive sections was equal to the thickness of the saw. The sections were taken from the mid-diaphysis and distal shaft. The thickness of the cross sections was reduced to about lOOum by grinding on #600 sandpaper. The material was not wet at the time of sanding (wet specimens tended to tear very easily).

The specimens were cleaned with an enzyme detergent powder (~ 1g Terg-A-Zyme, Alconox inc. and ~ 20ml of distilled water). The mixture was placed with the specimens in plastic containers and was not stirred or shaken but placed in an oven (Boekel Industries Inc. model 107800).

The numbers from the original bottles were assigned to the plastic containers as the specimens were placed inside them. The containers and their specimens remained in the oven at 30-50 degrees C for a week with one change of the detergent and water mixture during that time. Later the specimens were placed in an ultrasonic cleaner (Branson) in denatured ethyl alcohol for 15 seconds.

After cleaning, the specimens were removed and submerged in their mounting medium (DPX, Methanol Electro Pure or Gel Mount tm Aqueous mounting medium, Biomeda). They were ultrasonicated again for 15-20 seconds and then placed on a slide (Fisher brand 3"X1" and 1.2mm thick plain precleaned microscope) and covered with a cover glass (Fischer Finest 22X22mm, thickness No.1). Each specimen was mounted separately on a slide and each slide was given the same number that had been transferred from the original bottles onto the enzyme cleaning bottles. Better results were obtained using the Gel Mount.

An excess of bubbles could be move away with a bent tipped dissecting probe. The glass cover was place at a 45 degree angle and then pressed to further distance bubbles from the section. Weights (washers) were placed on the prepared slide on top of square cut cubes of grinding stone.

Photography

Polarized microscopy makes it possible to determine the orientation of the lamellae and crystallites. "It has been used to characterize the organization and orientation of lamellae within primary and secondary osteons, interstitial lamellar and circumferential lamellar regions" (Boyde and Riggs 90). Using sheets of circularly polarizing filter consisting of Polaroid filter sheet bonded to a quarter wavelength retarding material whose vibration direction is fixed at 45 degrees to that of the polarizing

18

material. Each filter may be rotated to pick up the desired colorless image that reveals the bright and dark lamellae. "One sheet is required for right and left, (or clockwise and counterclockwise) circularly polarized light (CPL), appropriately crossed" (Boyde and Riggs 90).

The mid-diaphysis and distal sections prepared from each femur were photographed using Ektar (100 ASA) color print film. The photographs were taken with an Olympus 35mm camera and Olympus binocular microscope. The camera was set on automatic. Each photograph contained a scale so that measurement could be directly taken from the photo image. The scale was later measured, converted and fed into the Micro-Plan II (Donsanto Corporation) program. The focus of this CPL study was to examine the narrow seams of collagen laid down over the experimental period (Figures 1-20).

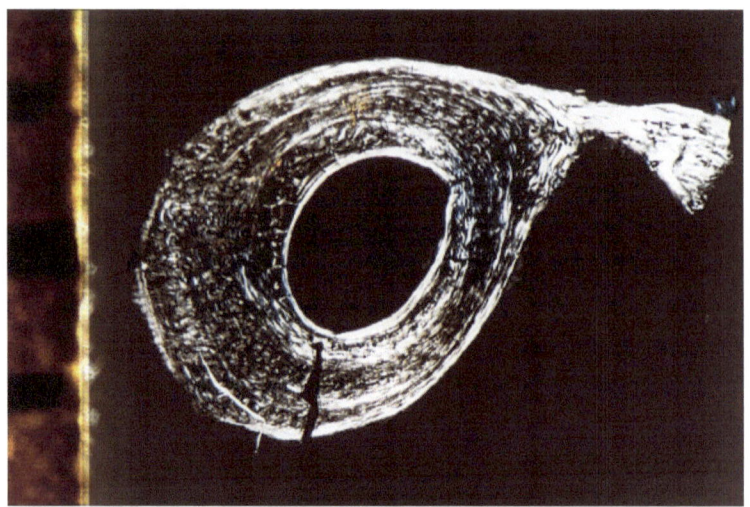

Figure No. 1, CPL photo, lOOum thick mid-diaphysis section of rat femur
Specimen #2, Left is anterior, Bottom is medial.

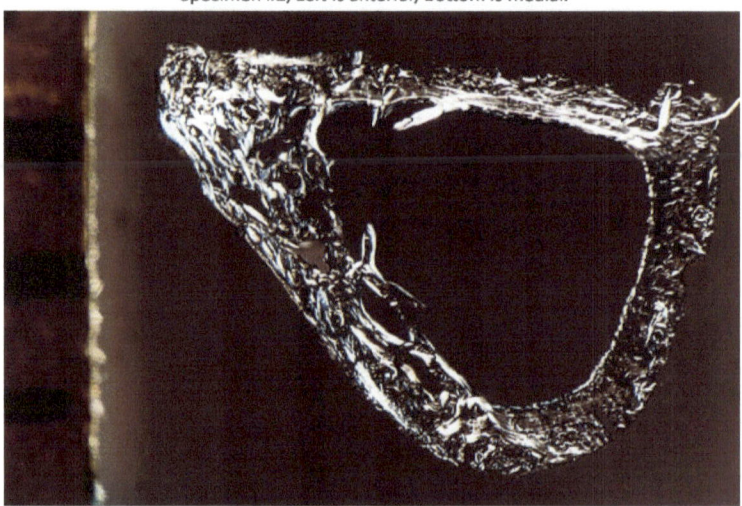

Figure No. 2, CPL photo, lOOum thick distal section of rat femur
Specimen #2, Left is anterior, Bottom is medial.

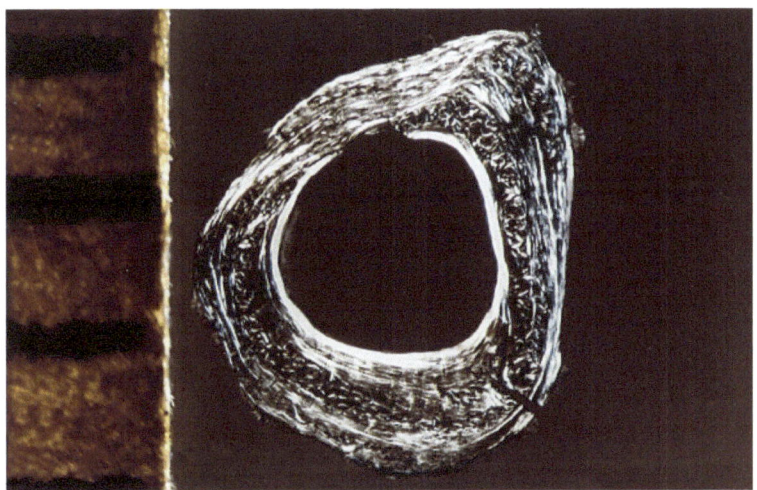

Figure No. 3, CPL photo, lOOum thick mid-diaphysis section of rat femur
Specimen #3, Bottom left is anterior, Top left is medial.

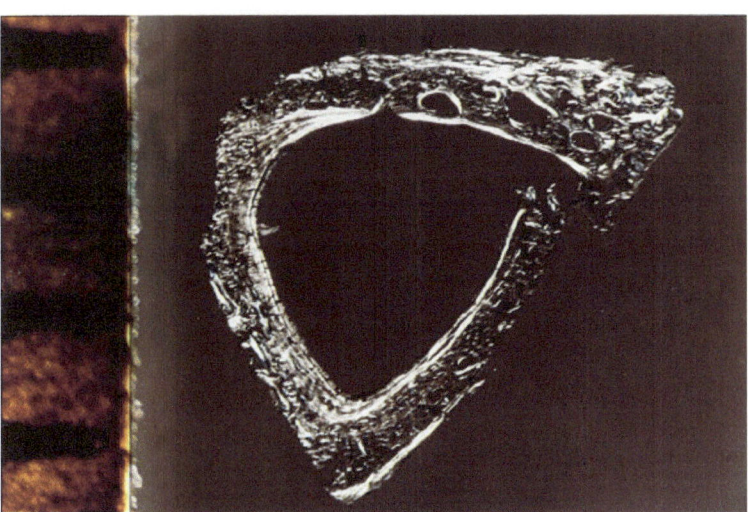

Figure No.4, CPL photo, lOOum thick distal section oi rat femur
Specimen #3, Top right is anterior, Top left is medial.

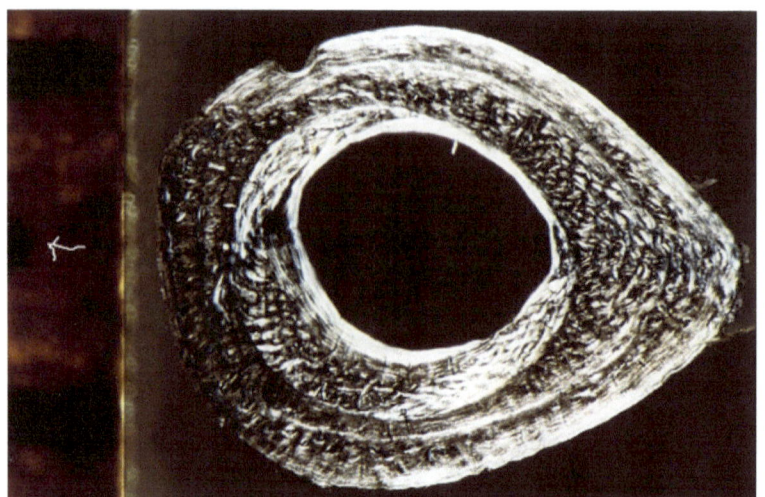

Figure No. 5, CPL photo, lOOum thick mid-diaphysis section of rat femur
Specimen #4, Left is anterior, bottom is lateral.

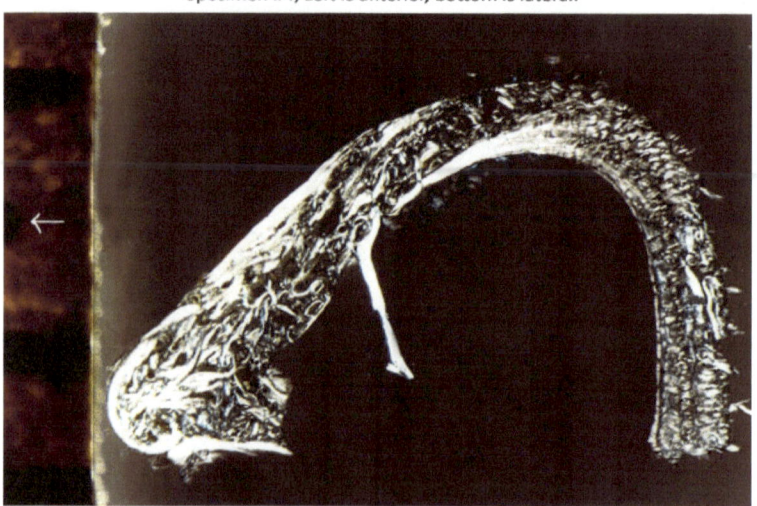

Figure No. 6, CPL photo, lOOum thick distal section of rat femur
Specimen #4, Left is anterior, Bottom is lateral.

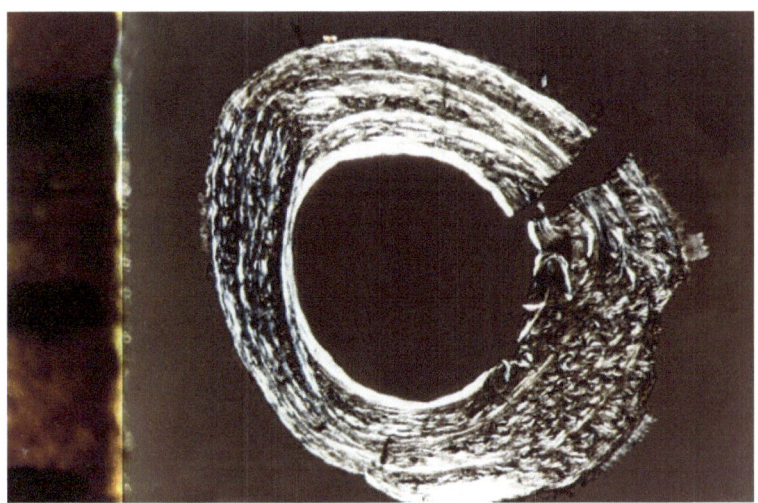

Figure No. 7, CPL photo, lOOum thick mid-diaphysis section of rat femur
Specimen #7, Top left is posterior, bottom left is medial

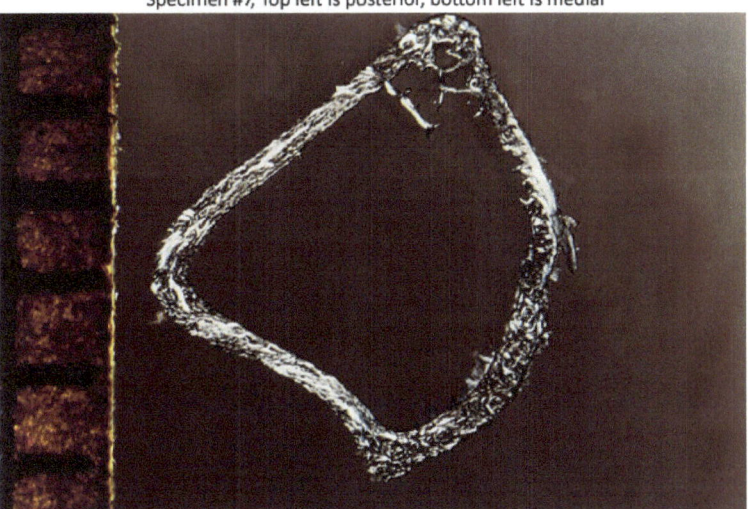

Figure No. 8, CPL photo, lOOum thick distal section of rat femur
Specimen #7, Top left is lateral, Bottom left is posterior.

Figure No. 9, CPL photo, lOOum thick mid-diaphysis section of rat femur
Specimen #8, Top left is anterior, Bottom left is lateral.

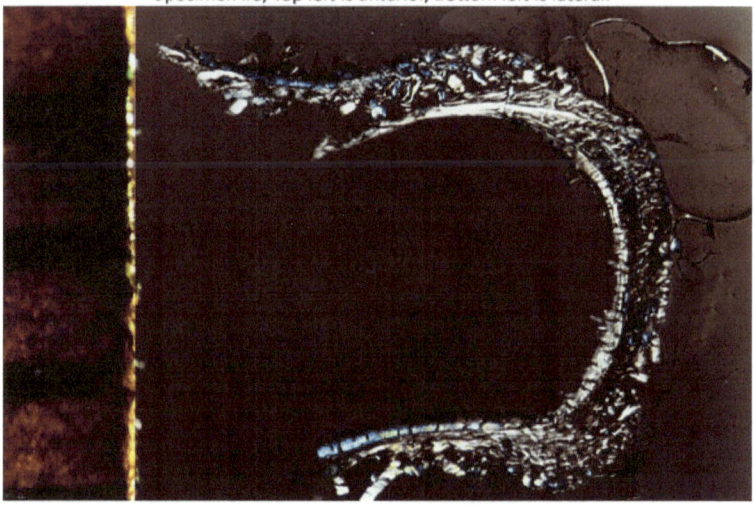

Figure No. 10, CPL photo, lOOum thick distal section of rat femur
Specimen #8, Right is posterior, Top is medial.

24

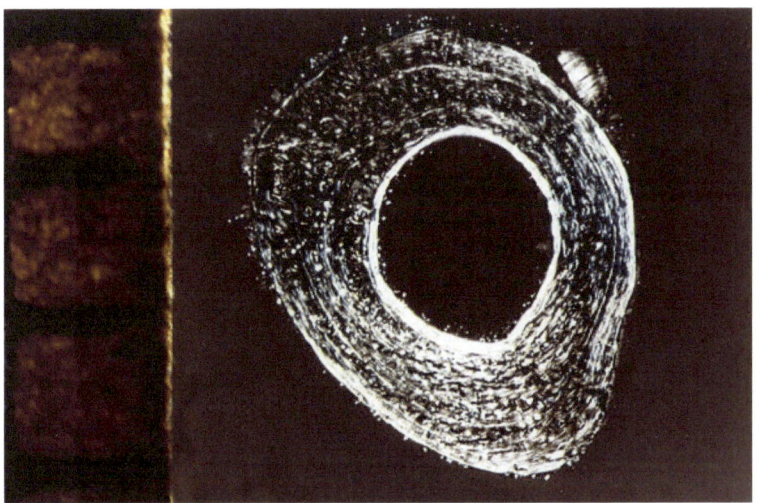

Figure No. 11, CPL photo, lOOum thick mid-diaphysis section of rat femur
Specimen #12, Left is lateral, Bottom is posterior.

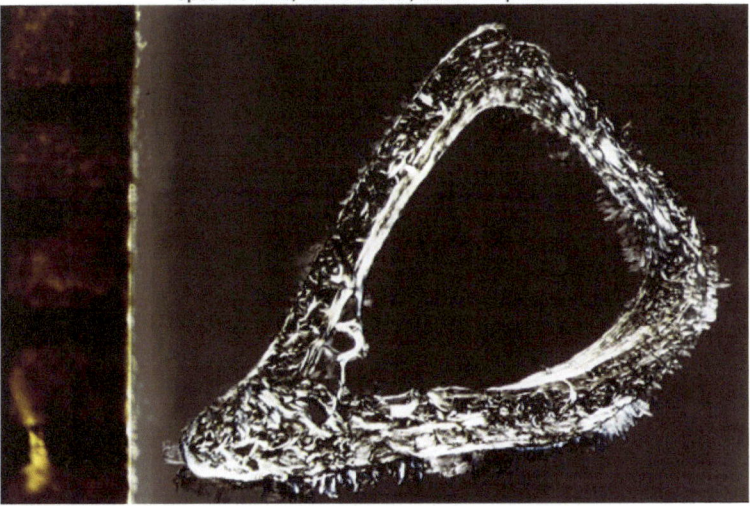

Figure No. 12, CPL photo, lOOum thick distal section 01 femur
Specimen #12, Left is anterior, Bottom is medial.

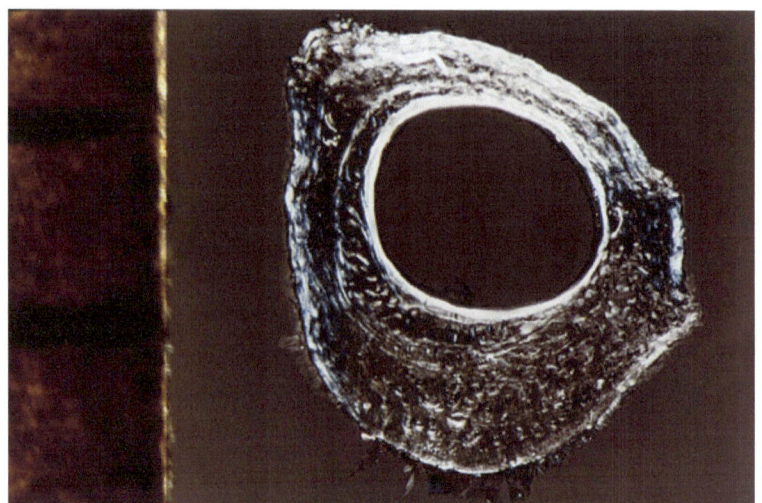

Figure No. 13, CPL photo, lOOum thick mid-diaphysis section of rat femur
Specimen #14, Bottom left is lateral, Bottom right is anterior.

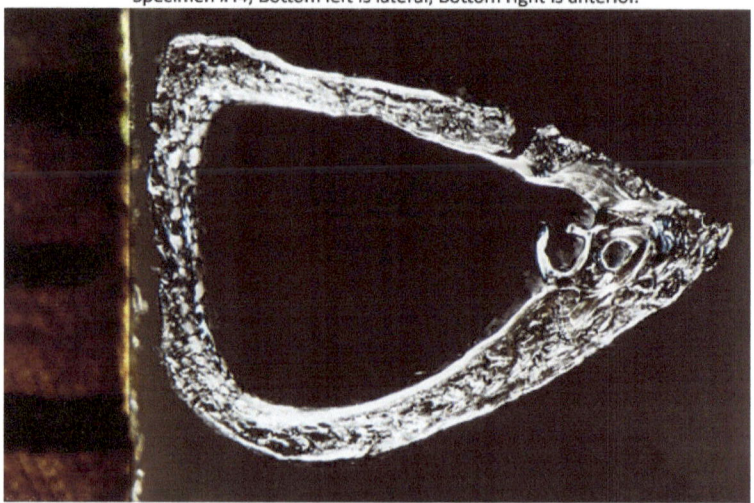

Figure No. 14, CPL photo, lOOum thick mid-diaphysis section of rat femur
Specimen #14, Left is posterior, Bottom is medial.

Figure No. 15, CPL photo, lOOum thick mid-diaphysis section of rat femur
Specimen #15, Left is medial, Bottom is posterior.

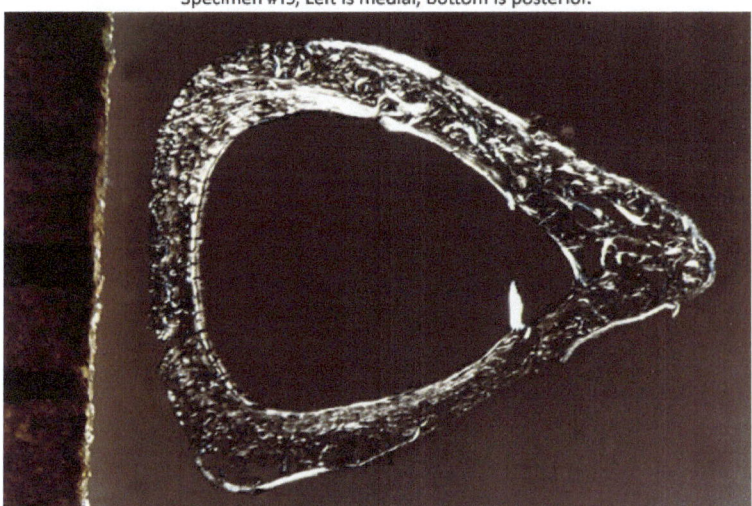

Figure No. 16, CPL photo, lOOum thick distal section of rat femur
Specimen #15 Left is posterior, Bottom is lateral.

Figure No.17, CPL photo,lOOum thick mid-diaphysis section of rat femur
Specimen #17, Left is medial, Bottom is posterior.

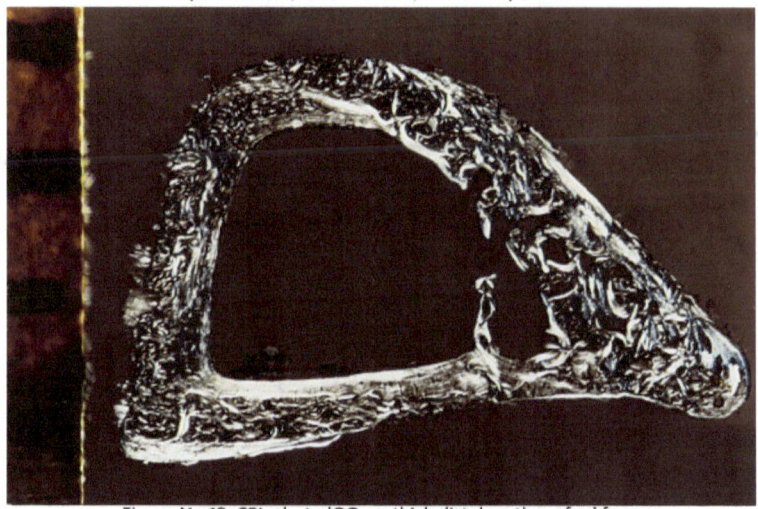

Figure No.18, CPL photo,lOOum thick distal section of ral femur
Specimen #17, Left is posterior, Bottom is lateral.

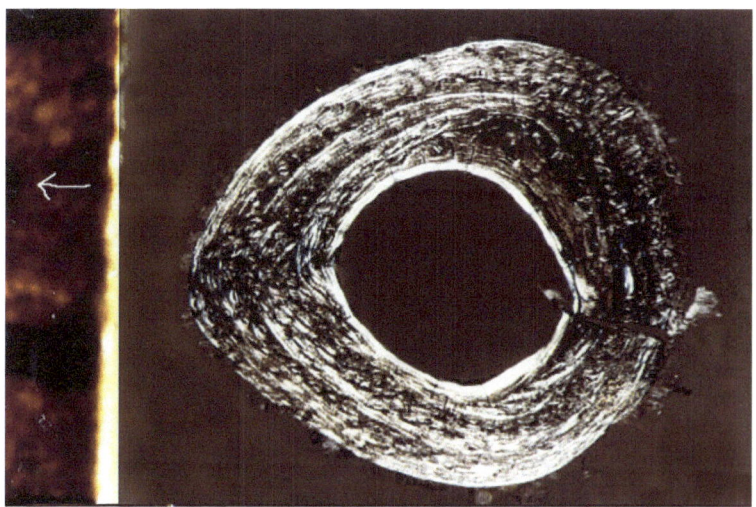

Figure No. 19, CPL photo, lOOum thick mid-diaphysis section of rat femur
Specimen #20, Left is posterior, Bottom is medial.

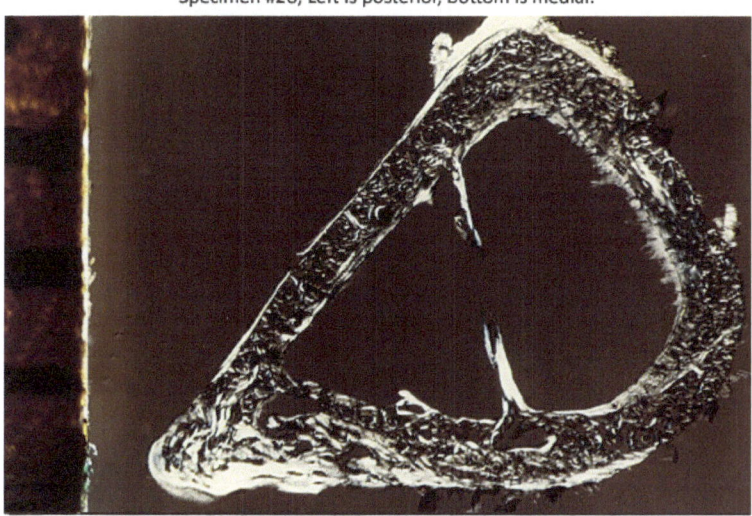

Figure No. 20, CPL photo, lOOum thick distal section of rat femur
Specimen #20, Left is anterior, Bottom is medial.

Micro-Measurements

Micro-Plan II (Donsanto Corporation) was employed to calculate the area of each prepared specimen (Table #1). This semiautomatic apparatus is designed for the acquisition and computation of geometric data, and its essential features are a digitizer tablet, a computer and a monitor. Bio-Image Analysis Version 1.0 (Spectra Services, 1992) was used with a video micrometer to measure widths of mid-diaphysis and distal slices and their cortices (Table #2). Observations and measurements of Haversian systems were also made (Table #3). All of the calculations are represented in schematic drawings for easy assessment (Figures 21-40).

TABLE #1.   Bio-Image Analysis 1.0 ; Data
MD=mid-diaphysis, D=distal, C(A)=cranial (anterior),
P=posterior, L=lateral, M=medial

Measurements in Microns (um)

No.	Spec	C(A)	P	L	M	A-P	L-M
		Rat Femur Cortex Thickness				Bone Width	
1.	#2MD	5440.81	6120.38	4483.87	3825.36	12032.57	8804.83
2.	#2D	5700.19	1980.37	1445.45	2472.90	19057.69	10764.50
3.	#3MD	2843.27	3770.54	1919.14	1459.68	13191.90	10984.61
4.	#3D	3821.94	1643.98	1513.57	1918.20	15306.39	11768.42
5.	#4MD	4194.76	5645.56	4202.01	3125.26	17462.30	14045.93
6.	#4D	4207.49	1609.88	0000.00	1792.09	16614.71	*9583.41
7.	#7MD	3188.64	2550.54	2735.02	1654.70	11261.86	9309.39
8.	#7D	951.01	855.61	923.64	1385.27	14516.18	13728.88
9.	#8MD	4207.68	4377.06	3087.77	3390.32	14852.91	12329.18
10.	#8D	0000.00	1851.78	1882.15	2420.70	00000.00	12468.03
11.	#12MD	3288.65	3594.90	2461.59	2078.18	11801.74	9134.94
12.	#12D	5013.80	1721.81	1745.88	2120.46	18424.14	12006.96
13.	#14MD	2659.59	3180.15	1136.94	3177.63	11049.79	9252.70
14.	#14D	3145.48	937.23	1008.87	1219.51	13053.64	8578.01
15.	#15MD	3494.68	3905.26	3095.06	2372.00	11760.94	9307.80
16.	#15D	4073.83	1451.61	1717.97	2126.51	16308.65	12438.09
17.	#17MD	3480.38	3964.85	2745.23	2138.15	11945.65	8755.22
18.	#17D	5986.48	1718.85	1614.92	2030.59	17564.18	10810.68
19.	#20MD	2936.98	3394.06	2580.65	2217.74	11296.15	9477.30
20.	#20D	5503.77	1602.85	1758.08	2091.87	18952.42	12661.55

* estimated

TABLE #2   Micro-Plan II : Data
Rat Femur
No. Specimen   Mean area

No.	Specimen	Midshaft	Distal	
1.	#2	4.814mm	5.504mm	
2.	#3	4.836mm	4.578mm	
3.	#4	5.054mm	5.714mm*	
4.	#7	3.468mm	5.559mm	
5.	#8	3.981mm	0.000	1.350mm post. area
6.	#12	4.602mm	6.364mm	1.372mm post. area
7.	#14	2.486mm	2.842mm	
8.	#15	4.982mm	5.894mm	
9.	#17	4.752mm	5.804mm	
10.	#20	5.759mm	8.100mm	

TABLE #3.   Bio-Image Analysis 1.0 : Data
Rat Femur
No. Specimen   Sample Haversion System Widths

No.	Specimen	Midshaft	Distal	Dark specimen
1.	#2	200-300um	450-720um	X
2.	#3	550-750um	200-400um	X
3.	#4	370-520um	300-700um	
4.	#7	320-510um	260-376um	
5.	#8	800-1080um	350um	X
6.	#12	500-840um	690-880um	X
7.	#14	490-730um	250-320um	
8.	#15	500-660um	500-970um	X
9.	#17	260-390um	330-650um	X
10.	#20	650-1750um	420-740um	X

32

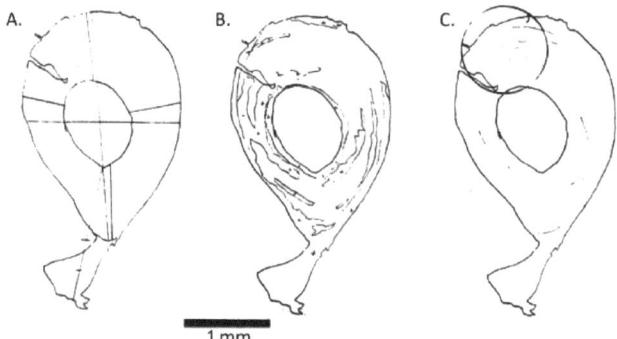

Figure No. 21, Schematic of CPL photo, Mid-diaphysis section of rat femur, Specimen #2, A. Cortex and diameter measurement locations, B. Bright lamellae areas, C. Lines = Haversian areas, Haversian system width samples taken from circled area. Left is medial, Top is anterior.

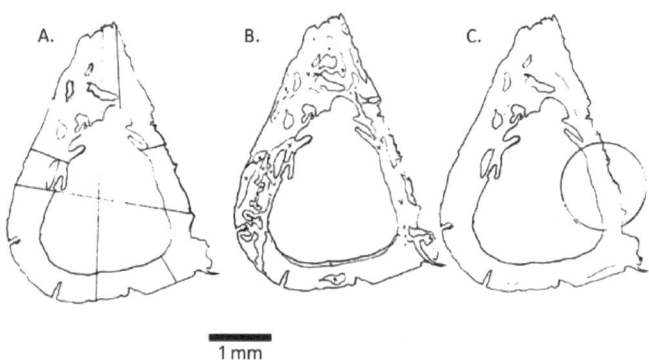

Figure No. 22, Schematic of CPL photo, Distal section of rat femur, Specimen #2, A. Cortex and diameter measurement locations, B. Bright lamellae areas, C. Lines = Haversian areas, Haversian system width samples taken from circled area. Left is medial, Top is anterior.

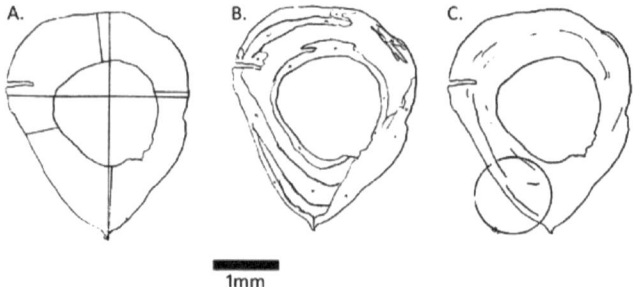

Figure No. 23, Schematic of CPL photo, Mid-diaphysis section of rat femur, Specimen #3, A. Cortex and diameter measurement locations, B. Bright lamellae areas, C. Lines = Haversian areas, Haversian system width samples taken from circled area. Left is lateral, Top is anterior.

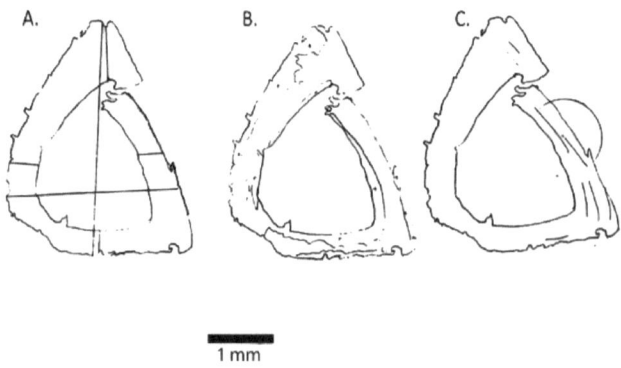

Figure No. 24, Schematic of CPL photo, Distal section of rat femur, Specimen #3, A. Cortex and diameter measurement locations, B. Bright lamellae areas, C. Lines = Haversian areas, Haversian system width samples taken from circled area. Left is medial, Top is anterior.

34

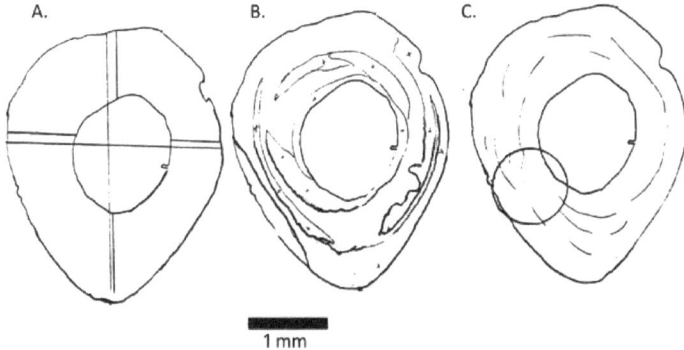

Figure No. 25, Schematic of CPL photo, Mid-diaphysis section of rat femur, Specimen #4, A. Cortex and diameter measurement locations, B. Bright lamellae areas, C. Lines = Haversian areas, Haversian system width samples taken from circled area. Left is lateral, Top is anterior.

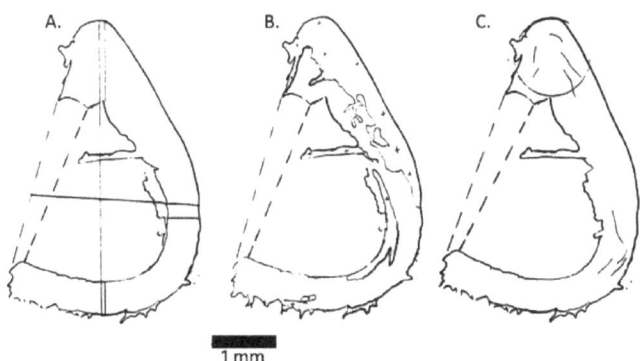

Figure No. 26, Schematic of CPL photo, Distal section of rat femur, Specimen #4, A. Cortex and diameter measurement locations, B. Bright lamellae areas, C. Lines = Haversian areas, Haversian system width samples taken from circled area. Left is lateral, Top is anterior.

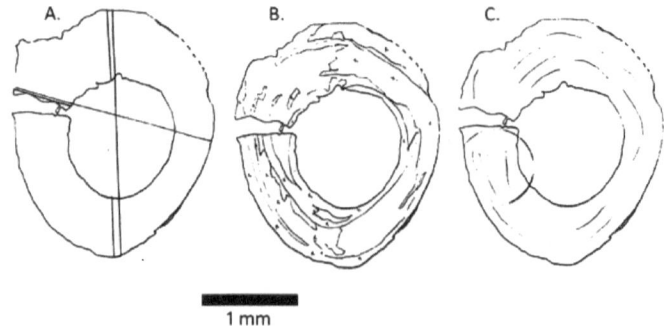

1 mm

Figure No. 27, Schematic of CPL photo, Mid-diaphysis section of rat femur, Specimen #7, A. Cortex and diameter measurement locations, B. Bright lamellae areas, C. Lines = Haversian areas, Haversian system width samples taken from circled area. Left is lateral, Top is anterior.

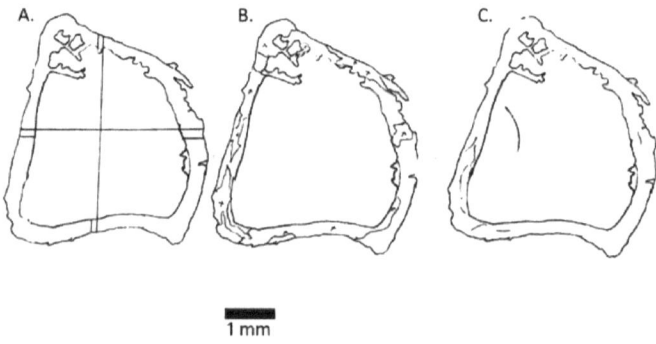

1 mm

Figure No. 28, Schematic of CPL photo, Distal section of rat femur, Specimen #7, A. Cortex and diameter measurement locations, B. Bright lamellae areas, C. Lines = Haversian areas, Haversian system width samples taken from circled area. Left is lateral, Top is anterior.

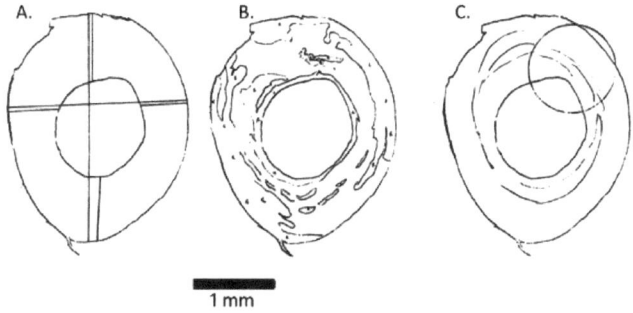

1 mm

Figure No. 29, Schematic of CPL photo, Mid-diaphysis section of rat femur, Specimen #8, A. Cortex and diameter measurement locations, B. Bright lamellae areas, C. Lines = Haversian areas, Haversian system width samples taken from circled area. Left is lateral, Top is anterior.

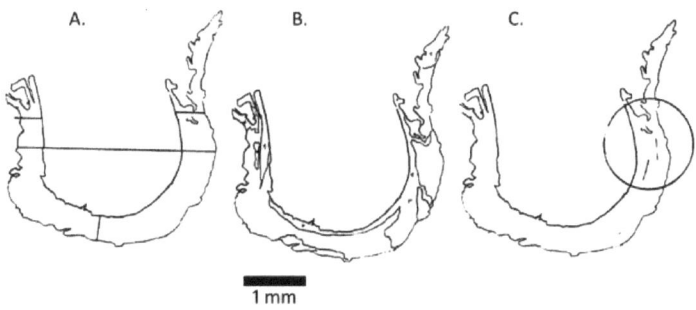

1 mm

Figure No. 30, Schematic of CPL photo, Distal section of rat femur, Specimen #8, A. Cortex and diameter measurement locations, B. Bright lamellae areas, C. Lines = Haversian areas, Haversian system width samples taken from circled area. Left is lateral, Bottom is posterior.

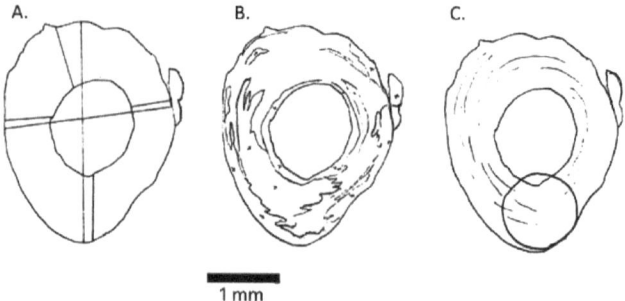

Figure No. 31, Schematic of CPL photo, Mid-diaphysis section of rat femur, Specimen #12, A. Cortex and diameter measurement locations, B. Bright lamellae areas, C. Lines = Haversian areas, Haversian system width samples taken from circled area. Left is lateral, Top is anterior.

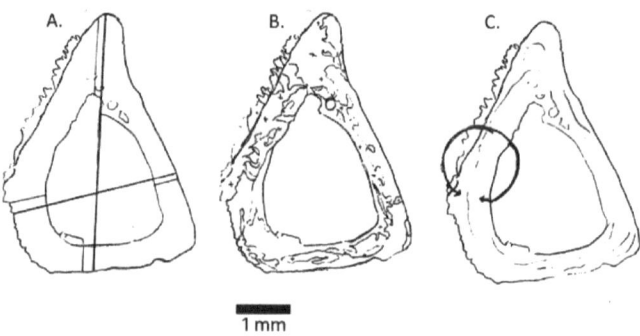

Figure No. 32, Schematic of CPL photo, Distal section of rat femur, Specimen #12, A. Cortex and diameter measurement locations, B. Bright lamellae areas, C. Lines = Haversian areas, Haversian system width samples taken from circled area. Left is medial, Top is anterior.

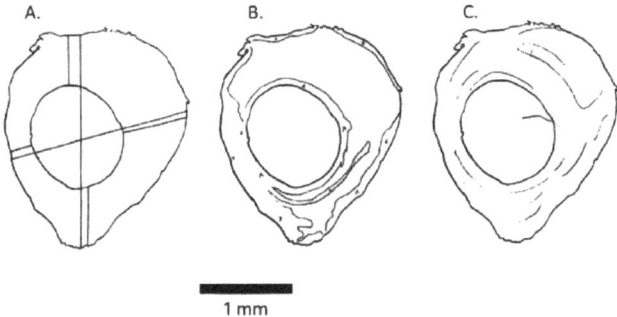

1 mm

Figure No. 33, Schematic of CPL photo, Mid-diaphysis section of rat femur, Specimen #14, A. Cortex and diameter measurement locations, B. Bright lamellae areas, C. Lines = Haversian areas, Haversian system width samples taken from circled area. Left is medial, Top is anterior.

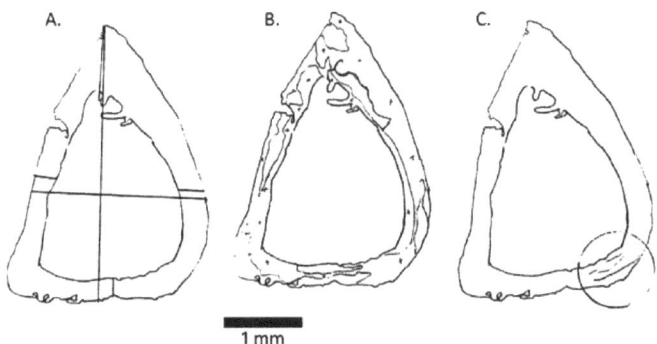

1 mm

Figure No. 34, Schematic of CPL photo, Distal section of rat femur, Specimen #14, A. Cortex and diameter measurement locations, B. Bright lamellae areas, C. Lines = Haversian areas, Haversian system width samples taken from circled area. Left is lateral, Top is anterior.

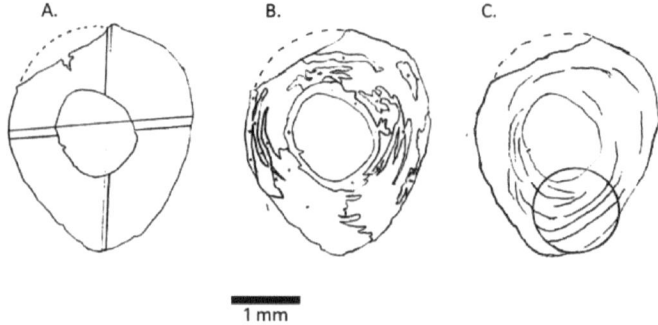

1 mm

Figure No. 35, Schematic of CPL photo, Mid-diaphysis section of rat femur, Specimen #15, A. Cortex and diameter measurement locations, B. Bright lamellae areas, C. Lines = Haversian areas, Haversian system width samples taken from circled area. Left is medial, Top is anterior.

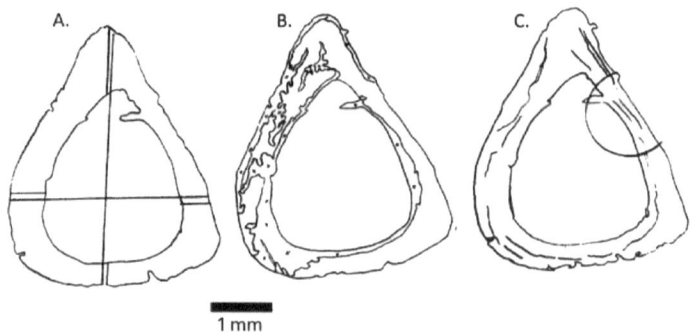

1 mm

Figure No. 36, Schematic of CPL photo, Distal section of rat femur, Specimen #15, A. Cortex and diameter measurement locations, B. Bright lamellae areas, C. Lines = Haversian areas, Haversian system width samples taken from circled area. Left is medial, Top is anterior.

40

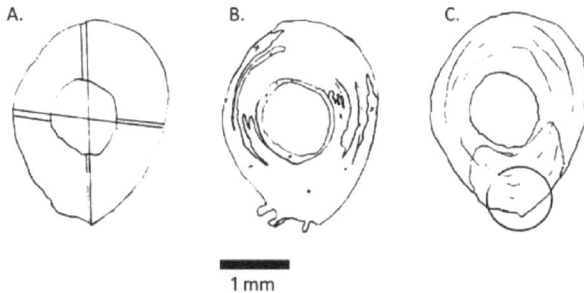

1 mm

Figure No. 37, Schematic of CPL photo, Mid-diaphysis section of rat femur, Specimen #17, A. Cortex and diameter measurement locations, B. Bright lamellae areas, C. Lines = Haversian areas, Haversian system width samples taken from circled area. Left is medial, Top is anterior.

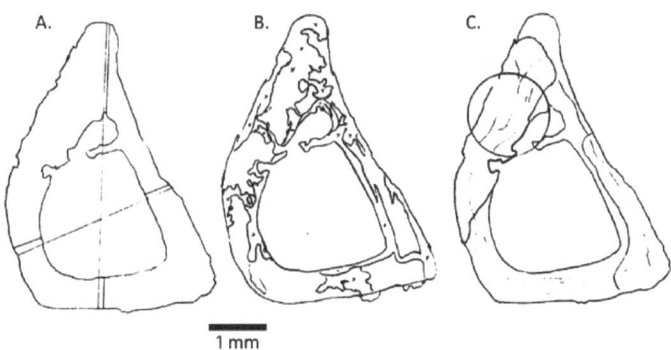

1 mm

Figure No. 38, Schematic of CPL photo, Distal section of rat femur, Specimen #17, A. Cortex and diameter measurement locations, B. Bright lamellae areas, C. Lines = Haversian areas, Haversian system width samples taken from circled area. Left is medial, Top is anterior.

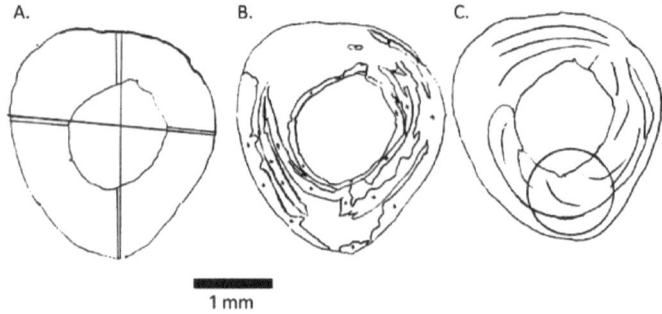

1 mm

Figure No. 39, Schematic of CPL photo, Mid-diaphysis section of rat femur, Specimen #20, A. Cortex and diameter measurement locations, B. Bright lamellae areas, C. Lines = Haversian areas, Haversian system width samples taken from circled area. Left is lateral, Top is anterior.

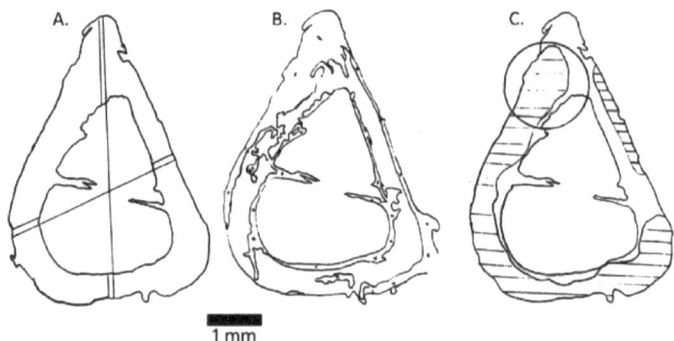

1 mm

Figure No. 40, Schematic of CPL photo, Distal section of rat femur, Specimen #20, A. Cortex and diameter measurement locations, B. Bright lamellae areas, C. Lines = Haversian areas, Haversian system width samples taken from circled area. Left is medial, Top is anterior.

Analysis

CPL: Mid-diaphysis

The circular polarized light reveals that all mid-diaphysis sections are characterized by circumferential lamellar bone layering the endosteal surface. Bright lamellae are found both medially and laterally with more intensity on the medial side. All bright lamellae found in the mid-shaft region are at the posterior end of the section.

CPL: Distal

In all complete specimens there is an observable deposit of lamellar bone layering the endosteal surface. The anterior and medial sides of these sections have bright lamellae still, as in the mid shaft sections, with more intensity on the medial side. Numbers 7, 10, 14, and 17 have bright lamellae on their lateral surfaces as well.

Results and Discussion

Past CPL studies of 100 micron sections of rat femur that have endured spaceflight have shown that this type of bone is generally bright posteriorly (composed of transversely oriented collagen) and dark anteriorly (composed of longitudinally oriented collagen). After analysis of the bone of the rats flown on board Cosmos 2044 Yamauchi et al. reported that osteoblastic genetic or epigenetic instructions over collagen alignment were not changed by weightlessness. Though there

43

may be alterations in bone mineral density as a result of space flight, the morphological character of collagen fiber orientation does not change.

It is now possible to report that much of the information collected during this study reveals that morphological character of the bone is continuous. The collagen fiber orientation stays intact as it did in the spaceflight bone. The most recognizable difference in a number of the specimens (marked as dark in Table #3 and done so during observations) is the concentration of Haversian systems and the range in their sizes. In Table #1 there is a visible link in the bone widths and the specimens with large Haversian systems in Table #3. The areas in Table #2 also seem to support a difference among the samples that occurs within the other tables. It is obvious when one looks at the specimens, marked (X) dark in Table #3, that they are different. The presence of these Haversian systems tells us that the rat was in an environment that activated the remodeling stimuli.

Conclusion

In comparing the spaceflight studies and the present one, at 2g, it is evident that the change in bone tissue is not on the morphological level but on the biochemical level. Changes in bone mineral density do not change the collagen orientation and architectural structure of the Rat femur.

References

Ascenzi, A.; Bannock, E.; Simkin, A.

1973 An approach to the mechanical properties of single osteonic lamellae J. Biomech 6: 227-235 et al., 1973;

Ascenzi, A.

1983 Microscopic dissection and isolation of bone constituents. In: Kunin, AS and Simmons, DJ (eds.): Skeletal Research, Vol2. New York: Academic Press, pp. 185-236.

Ascenzi, A.

1985 Mechanical hysteresis loops from single osteons: Technical devices and preliminary results. J. Biomech., 18:391-398.

Betram, J.E.A. and A.A. Biewener

1988 Bone Curvature: sacrificing strength for load predictability? In; J. Theor. Biology. 131:75-92.

Biewener, Andrew A., Shwart, S.M., Bertram, J.E.A.

1986 Bone modeling During Growth : Dynamic Strain Equilibrium in the Chick Tibiotarsus : Calcified Tissue International 39, New York Springer-Verlag Inc. pp. 390-395.

Boyde, A. and C.M. Riggs

1990 The Quantitative study of the Orientation of Collagen in Compact Bone slices. In: Bone, 11:35-39.

Boyde, A. and C.M. Riggs

1990 The Quantitative Study of the Orientation of
Collagen in Compact Bone Slices In: Bone, 11, 35-39.

Bromage, Timothy G.

1989a Ontogeny of the early hominid face Journal of Human
Evolution 18, pp. 751-773.

Bromage, Timothy G.

1989b Preferential Intracortical Remodeling and Growth of the
Macaque Mandible : Developmental Constraints on Growth
and Adaptation of Bone : Preprint, pp. 80 (not published).

Carando, S.; Portigliatti-Barbos, M.; Ascenzi, A.; Boyde, A.

1989 Orientation of collagen in human tibial and fibular shaft
and possible correlation with mechanical properties, In: Bone
10:139-142.

Cooper, R.R.; Milgram, J.W.; Robinson, R.A.

1966 Morphology of the Osteon. J. Bone Jt. Surg. 48A:
1239-1271.

Donaldson, C.L. et al.

1970 Effect of prolonged bed rest on bone metabolism :
Metabolism, 19, pp. 1071-1084.

Enlow, D.H. and Brown, S.O.

1957 A Comparative Histological Study of Fossil and Recent Bone Tissues. Parts II, II. Texas J. Sci. 9, pp. 186-214 and 10, pp. 187-230.

Garetto, L.P. et al.

1990 Preosteoblast production 55 hours after a 12.5-day spaceflight on Cosmos 1887 In: The FASEB Journal, Vol. 4, No. 1, pp. 24-28.

Harrison, G.A., Tanner, J.M., Pilbeam, D.R., Baker, P.T.

1989 Human Biology : An Introduction to Human Evolution, Variation, Growth and Adaptability, Oxford, England, Oxford University Press, pp. 568.

Jaworski Z.G. and Uhthoff H.K.,

1986 Reversibility of nontraumatic disuse osteoporosis during its active phase In: Bone 7, pp. 431 - 439.

Lanyon, L.E.

1984 Functional Strain as a Determinant for Bone Remodeling : Calcified Tissue International 36, New York, Springer-Verlag, pp. s56-s61.

Lanyon, L.E.

1986 Osteoporosis: Limitations of Strain-Related Bone Remodeling in Its Prevention In : Current Concepts of Bone Fragility, Springer-Verlag Berlin, pp. 195-196.

Lanyon, L.E.

1987 Functional Strain in Bone Tissue as an Objective, Controlling Stimulus for Adaptive Bone Remodeling : J. Biomechanics Vol. 20, No. 11/12, Great Britain, pp. 1083-1093.

Lanyon, L.E.

1992 Control of Bone Architecture by Functional Load Bearing In: Journal of Bone and Mineral Research Vol. 7., Supplement 2, Mary Ann Liebert, Inc., Pub. pp.s369-s375.

Mechanic, G.L. et al.

1990 Regional distribution of mineral and matrix in the femurs of rats flown on Cosmos 1887 biosatellite : The FASEB journal, Vol. 4, No. 1, pp. 34-40.

Portigliatti-Barbos, M.; , P.; Ascenzi, A.; Boyde, A.Bianco

1984 Collagen orientation in Compact Bone: II. Distribution of lamellae in the whole of the human femoral shaft with reference to its mechanical properties. In: Metab. Bone Dis. Rel. Res. 5;309-315; 1984.

Simmons, D.J., Grynpas, M.D., and Rosenberg, G.D.

1990 Maturation of bone and dentin matrices in rats flown on the Soviet biosatellite Cosmos 1887 : The FASEB journal, Vol. 4, No. 1, pp. 34-40.

Skerry, T.M., L. Bitensky, J. Chayen, and L.E. Lanyon

1988 Loading -Related Reorientation of Bone Proteoglycan in Vivo. Strain Memory in Bone Tissue? Journal of Orthopaedic Research, Raven Press Ltd., NY Orthopaedic Society pp. 547-551.

Smith, J. Maynard et al.

1985 Developmental Constraints and Evolution: The Quarterly Review of Biology Vol. 60, No. 3, pp. 265-287.

Tischler, Marc E, Christopher R. Kirby, U. of Arizona Health

1991 Space Travel (Biochemistry and Physiology) pp. 146-147. In: Encyclopedia of Human Biology Volume 7. Si-Zo, Renato Dulbecco Editor in Chief Salk Institute La Jolla CA, Academic Press, Inc. Harcourt Brace Jovanovich, Publishers

Tilton, F.E. et al.

1980 Long term follow up of Skylab bone demineralization: Aviat. Space Environ. Med. 51, pp. 1209-1212.

Uhthoff, H.K., Jaworski Z.F.G.

1978 Bone loss in response to long term immobilization: J. Bone Joint Surg. (Br) 60-B, pp. 420-429.

Vailas, A.C. et al.

1990 Effects of space flight on Humerus geometry, biomechanics and biochemistry.

Yamauchi, Mitsuo et al.

1991 Part II: Final Report For COSMOS 2044 EXPT K-701, Collagen, Cross-Links, Osteocalcin and Mineral Concentrations in the Diaphysis of Rat Femurs After 2 Weeks in Space.

About The Author

Adaptation of the Rat Femur to A Force of 2g led Wilson McCord to his work in theoretical and experimental biology on biological adaptation to stress environments and biological structures. His focus on the biological structure, mechanical system dynamics and reverse engineering of the neuromusculoskeletal system and bioinspired-design became clear and intensified after working as a Visiting Scholar and Research Associate in the Department of Zoology at the University of Cambridge in the summer of 2001.

Wilson McCord's multidisciplinary work has allowed him to make extraordinary contributions in biophysical and bioengineering research. Through theoretical and experimental research he has contributed to the understanding of; the biomechanical function and mechanisms of movement in insects and crustaceans as well as the morphology, maintenance, mechanical properties and regeneration of tissue structures within the neuromusculoskeletal system. His contributions include: bone regeneration experiments as a Research Support Specialist in the Department of Biomedical Engineering at Stony Brook and, as a contracted expert in functional biomechanical design and kinematics, highly significant contributions to a neurobiology and biorobotics project. Mr. McCord has worked as a research assistant in the biological anthropology laboratories in the Department of Anthropology at Hunter College of the City University of New York (CUNY), attended lectures at the Mount Sinai School of Medicine and studied at the American Museum of Natural History and the New York Botanical Garden. He holds a B.A. in

Art and Design, and an M.A. in Anthropology from Hunter College of the City University of New York.

WILSON MCCORD SCIENCE PUBLICATIONS

www.ingramcontent.com/pod-product-compliance
Lightning Source LLC
Chambersburg PA
CBHW041109180526
45172CB00001B/181